TRANSFORMING ANIMALS

TURNING INTO A FISH

by Tyler Gieseke

Cody Koala
An Imprint of Pop!
popbooksonline.com

abdobooks.com
Published by Pop!, a division of ABDO, PO Box 398166, Minneapolis, Minnesota 55439. Copyright ©2022 by Abdo Consulting Group, Inc. International copyrights reserved in all countries. No part of this book may be reproduced in any form without written permission from the publisher. Cody Koala™ is a trademark and logo of Pop!.

Printed in the United States of America, North Mankato, Minnesota

102021
012022

THIS BOOK CONTAINS RECYCLED MATERIALS

Cover Photo: iStockphoto
Interior Photos: Shutterstock Images, 1–7, 10–15, 19–20; wiljoj / Getty Images, 9; Krzysztof Winnik / Alamy Stock Photo, 16–17

Editor: Elizabeth Andrews
Series Designers: Laura Graphenteen, Victoria Bates

Library of Congress Control Number: 2021942284
Publisher's Cataloging-in-Publication Data
Names: Gieseke, Tyler, author.
Title: Turning into a fish / by Tyler Gieseke
Description: Minneapolis, Minnesota : Pop!, 2022 | Series: Transforming animals | Includes online resources and index.
Identifiers: ISBN 9781098241162 (lib. bdg.) | ISBN 9781098241865 (ebook)
Subjects: LCSH: Fishes--Juvenile literature. | Marine animals--Juvenile literature. | Animal life cycles--Juvenile literature. | Metamorphosis--Juvenile literature. | Animal Behavior--Juvenile literature.
Classification: DDC 571.876--dc23

Hello! My name is

Cody Koala

Pop open this book and you'll find QR codes like this one, loaded with information, so you can learn even more!

Scan this code* and others like it while you read, or visit the website below to make this book pop.

popbooksonline.com/turn-fish

*Scanning QR codes requires a web-enabled smart device with a QR code reader app and a camera.

Table of Contents

Chapter 1
Transforming Animals 4

Chapter 2
Larva and Fry 8

Chapter 3
Growth and Change 12

Chapter 4
Later Life 18

Making Connections 22
Glossary 23
Index 24
Online Resources 24

Chapter 1

Transforming Animals

There are many kinds of fish. Some are colorful. Others are different shapes.

All fish change as they get older. Some types change more than others.

Watch a video here!

Fish are **transforming** animals. They grow through five steps. The steps are egg, larva, fry, juvenile, and adult. Together, these steps make the fish life cycle.

Every fish begins life as an egg. Fish eggs are often small and sticky. Females can lay thousands of eggs.

> South American lungfish breathe water as young fish. After seven weeks, they start to breathe air!

Chapter 2

Larva and Fry

Fish eggs **hatch** after several days. When a fish hatches, it is called a larva. It has a **yolk sac** stuck to it. The sac gives the larva **nutrients** to grow.

Learn more here!

A larva is the second step in the fish life cycle. It does not eat yet.

When the larva finishes the yolk sac, it is called a fry. It can go find food. Many fish spend a few months as fry.

> A **lake sturgeon** larva is done with its yolk sac after nine to 18 days.

Chapter 3

Growth and Change

The fourth step in the fish life cycle is the juvenile. These fish often look like small **versions** of adult fish.

Complete an activity here!

adult

Juvenile **catfish** have a whitish color. They have dark spots on their fins. When they are adults, they will be darker.

Certain fish are juveniles for a long time. **Lake sturgeon** spend 15 to 20 years in this part of the life cycle. The step is shorter for other fish. Catfish spend five to eight years as juveniles.

Chapter 4

Later Life

Juvenile fish grow into adults. Adult fish can make new eggs. They won't change much for the rest of their lives.

Learn more here!

Some adult fish **spawn** every year. Others do this only once. Then, they die. With new eggs, the life cycle starts again! Fish are amazing **transforming** animals.

Making Connections

Text-to-Self

Which is your favorite step in the fish life cycle? Why?

Text-to-Text

What stories have you read about fish? Did they include fish at different steps in the life cycle? Why do you think they did or didn't?

Text-to-World

What is another transforming animal you know about? How is its life cycle similar to or different from a fish's?

Glossary

catfish – a fish with whiskers near its mouth.

hatch – to break out of an egg.

lake sturgeon – a big fish that lives in fresh water.

nutrient – a thing in food the body needs to grow.

spawn – to create many new eggs in the water.

transform – to change into a new shape.

version – a copy that is a little different from the first thing.

yolk sac – a bundle attached to young animals that holds nutrients.

Index

catfish, 15–16

color, 4, 15

eggs, 6–8, 18, 20–21

food, 8, 10

lake sturgeon, 11, 16

life cycle, 6, 11–12, 20–21

spawn, 21

Online Resources

popbooksonline.com

Thanks for reading this Cody Koala book!

Scan this code* and others like it in this book, or visit the website below to make this book pop!

popbooksonline.com/turn-fish

*Scanning QR codes requires a web-enabled smart device with a QR code reader app and a camera.